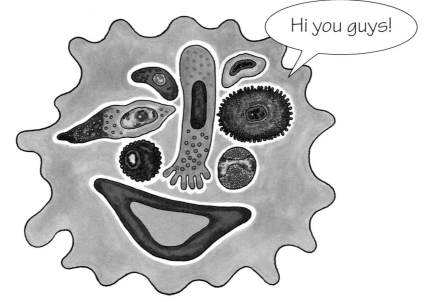

ENJOY YOUR CELLS

Fran Balkwill & Mic Rolph

Cold Spring Harbor Laboratory Press

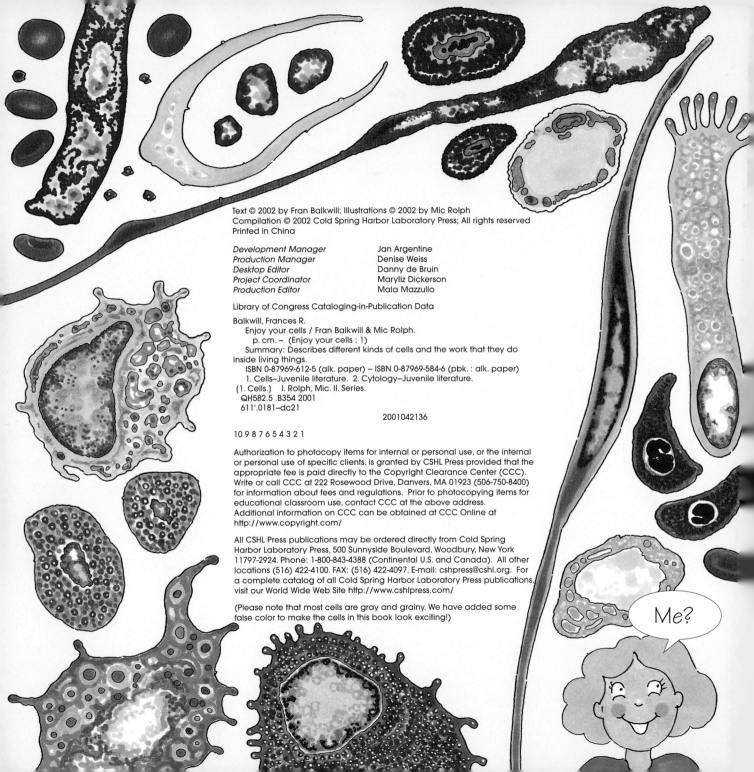

Development Manager   Jan Argentine
Production Manager   Denise Weiss
Desktop Editor   Danny de Bruin
Project Coordinator   Maryliz Dickerson
Production Editor   Mala Mazzullo

Library of Congress Cataloging-in-Publication Data

Balkwill, Frances R.
    Enjoy your cells / Fran Balkwill & Mic Rolph.
       p. cm. – (Enjoy your cells ; 1)
    Summary: Describes different kinds of cells and the work that they do
inside living things.
       ISBN 0-87969-612-5 (alk. paper) – ISBN 0-87969-584-6 (pbk. : alk. paper)
       1. Cells–Juvenile literature. 2. Cytology–Juvenile literature.
   (1. Cells.)   I. Rolph, Mic. II. Series.
    QH582.5 .B354 2001
    611'.0181–dc21
                                        2001042136

10 9 8 7 6 5 4 3 2 1

All CSHL Press publications may be ordered directly from Cold Spring
Harbor Laboratory Press, 500 Sunnyside Boulevard, Woodbury, New York
11797-2924. Phone: 1-800-843-4388 (Continental U.S. and Canada). All other
locations (516) 422-4100. FAX: (516) 422-4097. E-mail: cshpress@cshl.org. For
a complete catalog of all Cold Spring Harbor Laboratory Press publications,
visit our World Wide Web Site http://www.cshlpress.com/

(Please note that most cells are gray and grainy. We have added some
false color to make the cells in this book look exciting!)

Just beneath your skin there is an amazing, hidden world.

A world of living cells that work together to make you.

Cells are microscopic building blocks.

Millions and millions of cells come together to build your blood and brain and bones and muscles and skin; they grow your hair, your teeth, and your nails. Cells form your liver, lungs, heart and kidneys, and all the other bits inside.

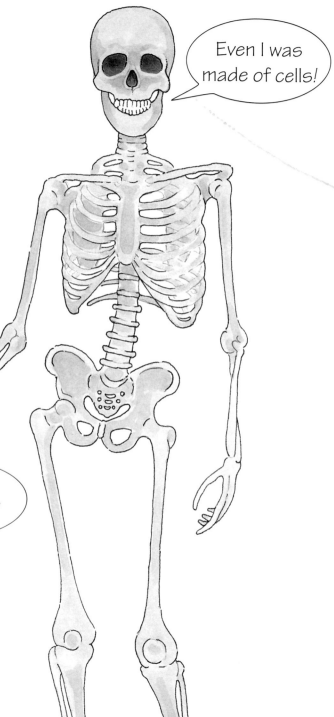

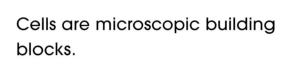

But what about other living things? Are they made of cells as well?

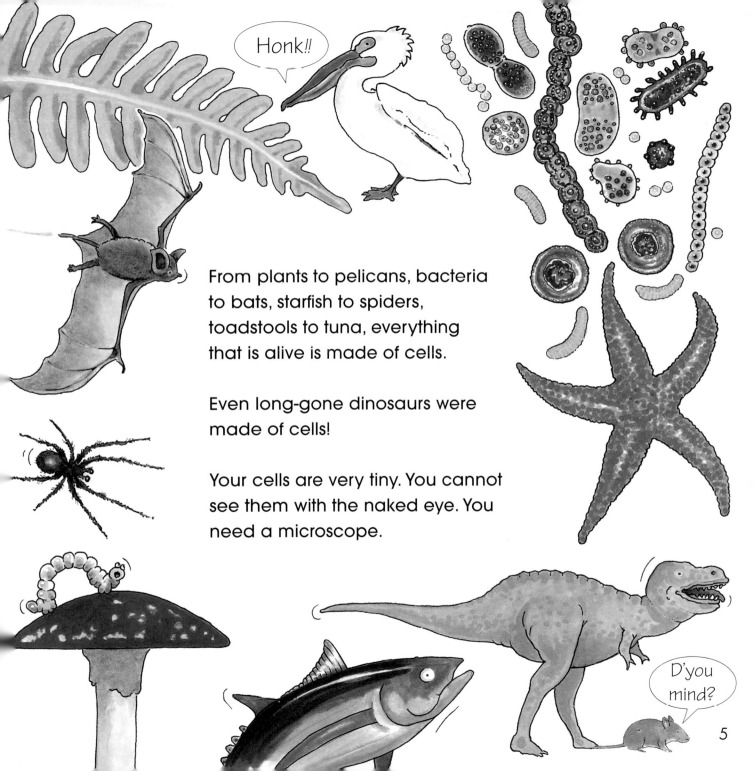

Honk!!

From plants to pelicans, bacteria to bats, starfish to spiders, toadstools to tuna, everything that is alive is made of cells.

Even long-gone dinosaurs were made of cells!

Your cells are very tiny. You cannot see them with the naked eye. You need a microscope.

D'you mind?

5

If you cut a very thin slice through a human cell that has been magnified ten thousand times by a powerful electron microscope, you see that it is quite complicated.

On the outside, there is a membrane, a bit like a soap bubble. This membrane encloses jelly-like cytoplasm (sigh-toe-plaz-um) which is crammed with many blob-like objects and tubes. There is also a nucleus (new-clee-us) that controls the cell.

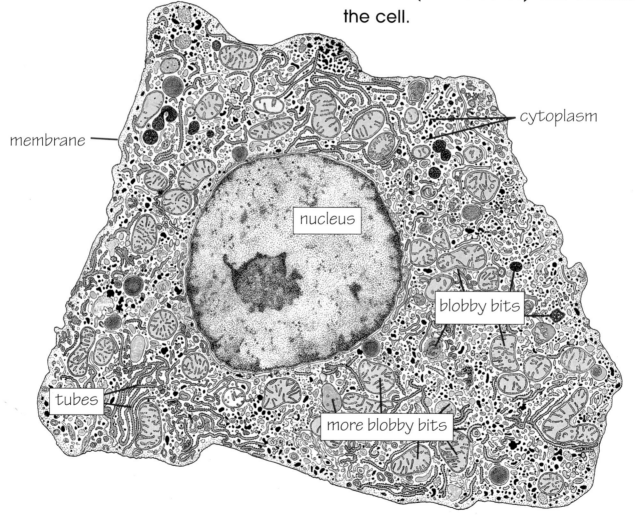

membrane

cytoplasm

nucleus

blobby bits

tubes

more blobby bits

These strange-looking bits inside the cell do all kinds of clever jobs; like turning food into energy and making important chemicals called proteins (pro-teens). This means that cells can work together to make you eat, sleep, think, and talk; they make you breathe, sing, run, and walk.

I'll never make PRESIDENT!!

You never know!

This amoeba has been magnified 3000 times.

Although many living things are made of millions of cells, some creatures, like an amoeba (am-ee-baa), exist as just one cell.

Not all the cells are the same size.

Most of the cells in your body are about 1/1000 inch wide. It would take 25 to cover the surface of a grain of sand.

The smallest type of cell is a bacterium (back-teer-ee-um) and, like the amoeba, it usually lives on its own. A bacterium is about 1/10,000 inch long. You could fit 2500 bacteria on a grain of sand.

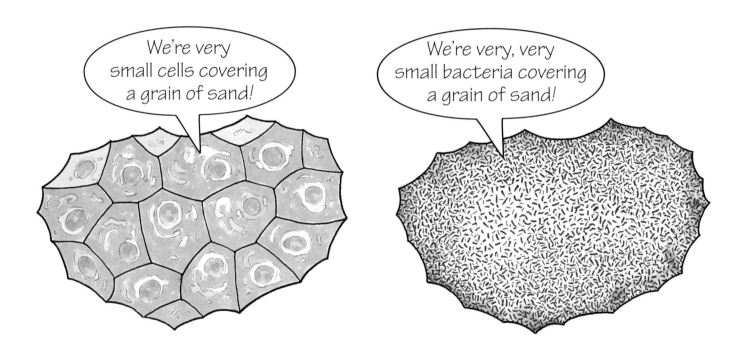

This is the size of a grain of sand.
Can you see it?

Here is another way to understand the size of your cells.

Look at your thumb. If each of the cells were ½ inch wide, your thumb would be as long as a baseball diamond, or as tall as a nine-story building.

9

We have already told you that cells work together to build living creatures.

So, how many human cells does it take to make your body?

Errrr... one hundred?

One million?

One hundred million?

NO!

You are made of about one hundred million million (100,000,000,000,000) tiny cells!

I thought I was made of **LATEX!**

That **is** a lot of cells!

But what about everything else? Do elephants and giant redwood trees have big cells? No—they just have more cells than you.

Even most dinosaur cells were the same size as yours. The biggest dinosaur ever known was probably made of about 1000 million million cells.

11

At your very beginning, nine months before you were born, there were just two very special cells.

There was a small cell with a wiggly tail called a sperm, and a larger pillowy egg cell.

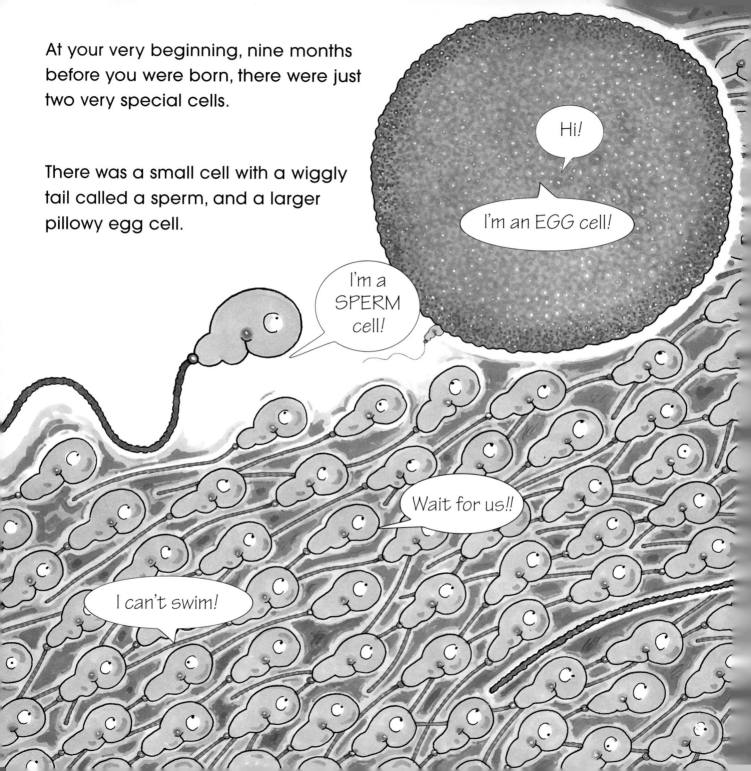

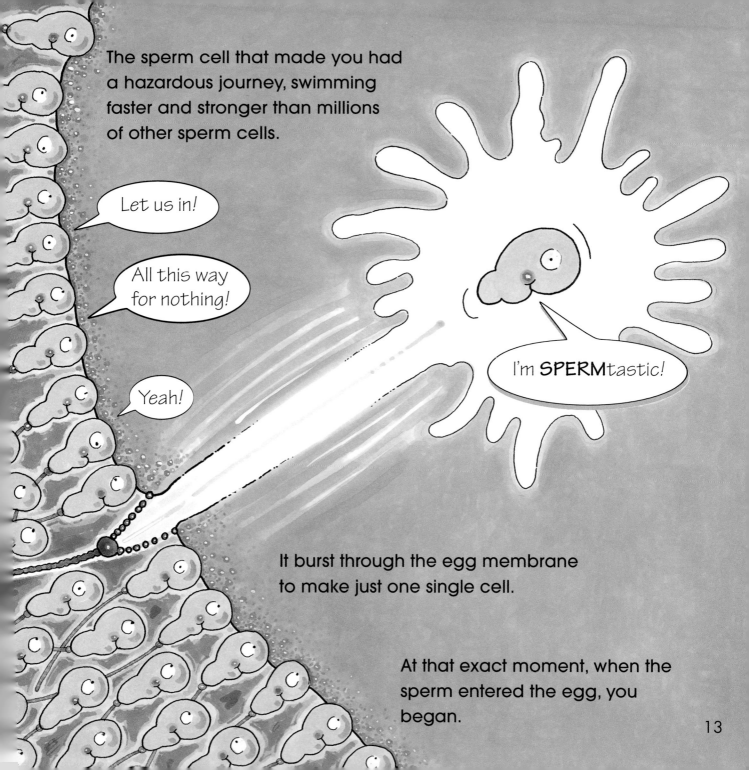

The sperm cell that made you had a hazardous journey, swimming faster and stronger than millions of other sperm cells.

Let us in!

All this way for nothing!

Yeah!

I'm **SPERM**tastic!

It burst through the egg membrane to make just one single cell.

At that exact moment, when the sperm entered the egg, you began.

13

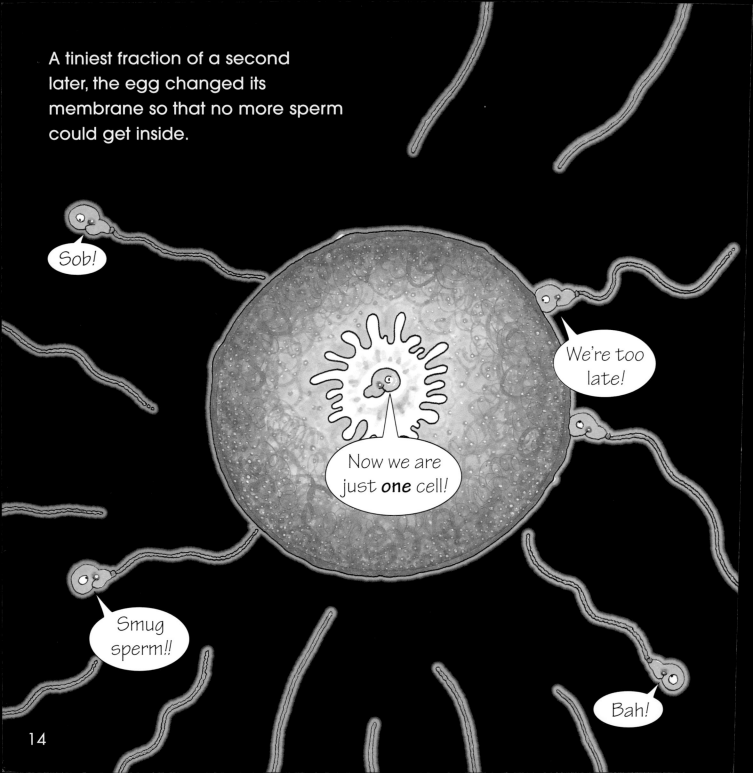

Within a few hours, dark shapes could be seen inside the cell. These were your chromosomes (kro-muh-sohmz).

They held all the information and instructions to make you, in a secret chemical code.

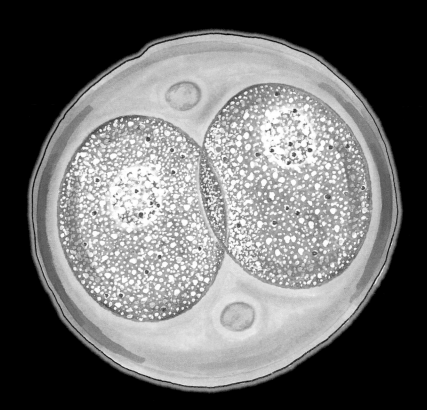

Very soon your first cell had split
into two cells.

You were on your way!

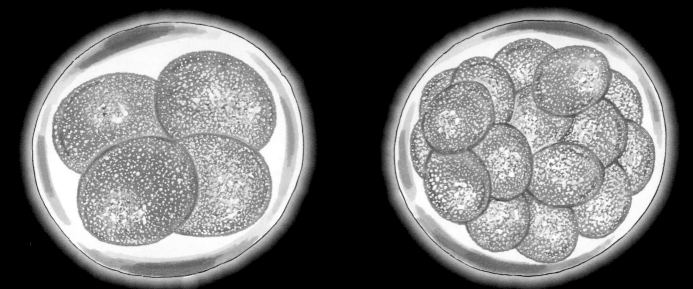

A few hours later your 2 cells had become 4 cells. Then 4 cells became 8 cells. Then 8 cells became 16 cells, 16 cells became 32 cells...

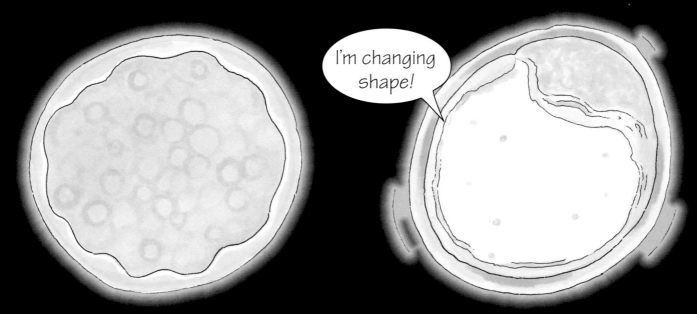

...and so on until you were an embryo (em-bree-oh) of about 200 cells. At this stage you were shaped a bit like a ball with a hollow center.

Your cells had already begun to look different from one another. They moved and divided in a precise order, and began to make all the parts of your body.

One month after the egg and sperm first met, you were about ¼ inch long. Your tiny body was safe and warm inside a sac of liquid. And still your cells kept making more cells. By 2 months, you were about 1½ inches long.

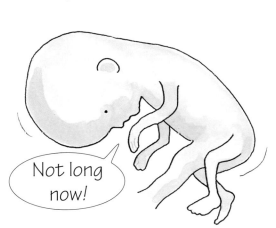

During the next 2 months your cells made many millions more cells. You began to use your arms and legs to explore your watery world. Still your cells made more cells and you kept on growing. About 9 months after the sperm cell met the egg cell... you were born.

And you still kept on growing by making more cells. You became taller, wider, fitter, stronger, and you are probably still growing now.

How do your cells know when to grow, and when to make more cells? How is all this controlled?

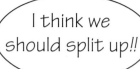

I think we should split up!!

Yep!

Sob!

Cells are controlled by a very special chemical called deoxyribonucleic acid (**D**ee-ahk-see-rye-boh-**N**ew-clay-ik **A**cid).

DNA is much easier to say.

DNA is the secret chemical code that was in the chromosomes of the first cell that was you. And before your first cell became two cells (and, indeed, before any of your cells became two cells) that DNA was copied. **How?**

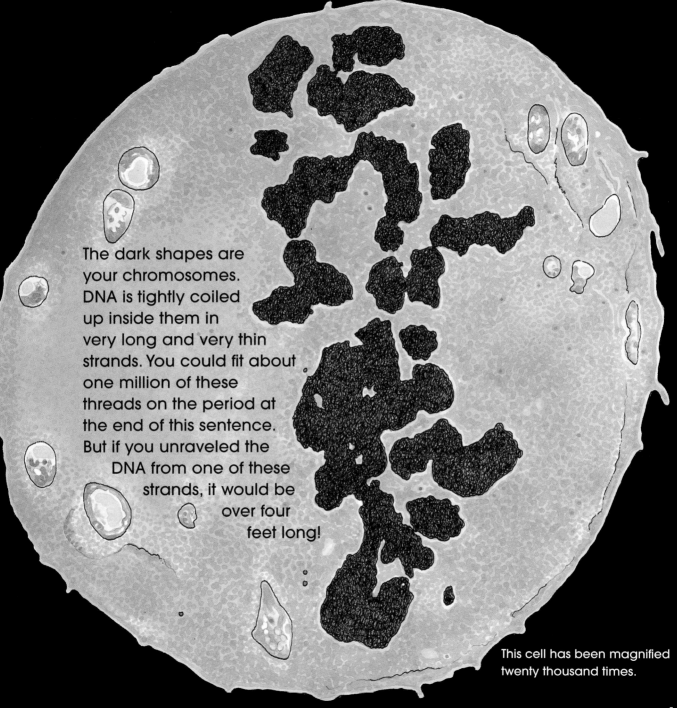

The dark shapes are your chromosomes. DNA is tightly coiled up inside them in very long and very thin strands. You could fit about one million of these threads on the period at the end of this sentence. But if you unraveled the DNA from one of these strands, it would be over four feet long!

This cell has been magnified twenty thousand times.

We have unraveled a part of the DNA thread from a cell and magnified it about 70 million times. Now you can see that it is made of two strands that twist around each other like a winding, twirling ladder. This shape is called a double helix.

The chemicals that make DNA are floating nearby and join up in a precise order.

This is how DNA is copied. But how does DNA make you?

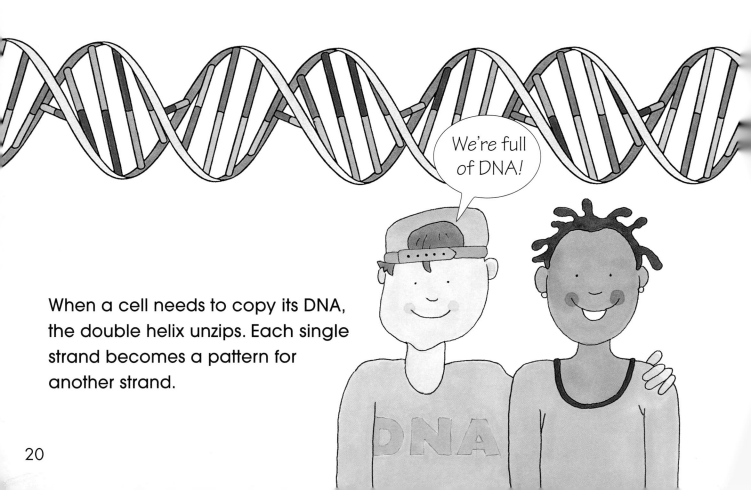

We're full of DNA!

When a cell needs to copy its DNA, the double helix unzips. Each single strand becomes a pattern for another strand.

20

Well, DNA is a code for making proteins. Your DNA plan contains about 30,000 different recipes for making proteins.

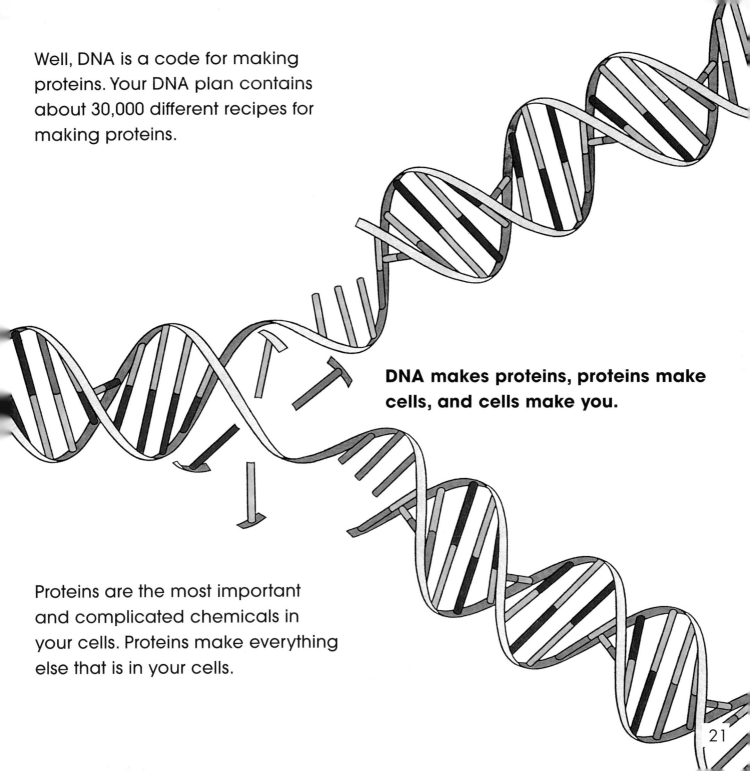

DNA makes proteins, proteins make cells, and cells make you.

Proteins are the most important and complicated chemicals in your cells. Proteins make everything else that is in your cells.

The protein message in your DNA code combines in many different ways, to make more than 200 types of cells. Let's meet some of them.

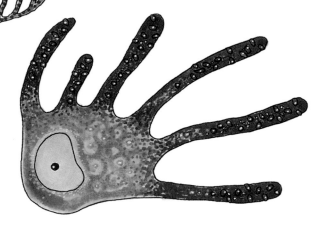

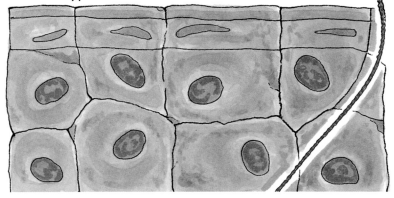

Skin cells keep the outside out and your insides in. They protect you from the sun and rain and cold.

Pigment cells give your skin its color. All of us have the same number of pigment cells in our skin. From pale yellow to black, the color of your skin depends on how much melanin is in your pigment cells.

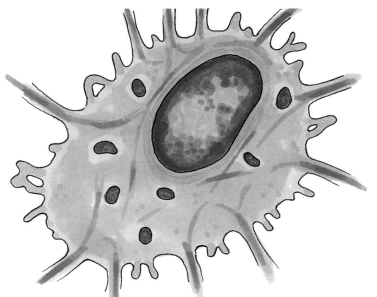

Hair cells make a tough protein called keratin—that's what hair is—pure keratin.

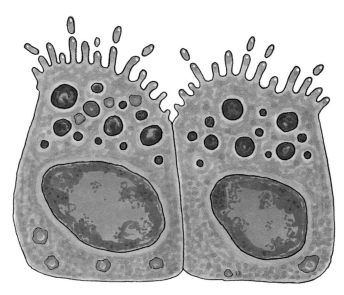

Gland cells in your skin make special oils. These oils stop your hair from becoming too brittle, and keep your skin nice and soft.

Fibroblasts (fye-broh-blasts) are "builder cells." They produce a sticky glue that keeps your cells together. Fibroblasts are good at repairing wounds when you cut yourself.

Muscle cells are stretchy cells. They can change their size and shape to move almost every part of your body.

Stick with me!

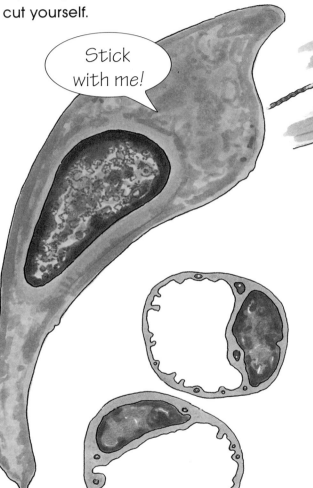

Nerve cells carry electrical messages. They monitor the world inside and outside your body, sending messages back to your brain. Nerve cells are the longest cells in your body—they may be only 1/1000 inch wide, but they can be 3 feet long.

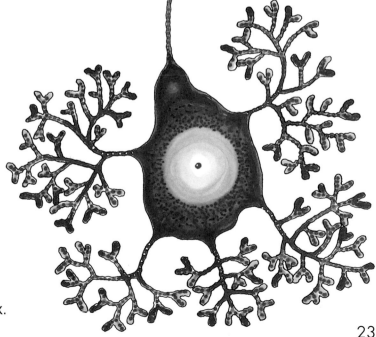

The **endothelial** (end-oh-theel-ee-all) **cells** that line your blood vessels are very thin. The tiniest blood vessels are just one cell thick.

23

Neutrophil (new-tro-fil) defender cells love to gobble up germs and destroy them with deadly chemical weapons.

Dendritic (den-drit-ick) cells capture bits of germs. Then they go to warn cells called lymphocytes that there is trouble ahead.

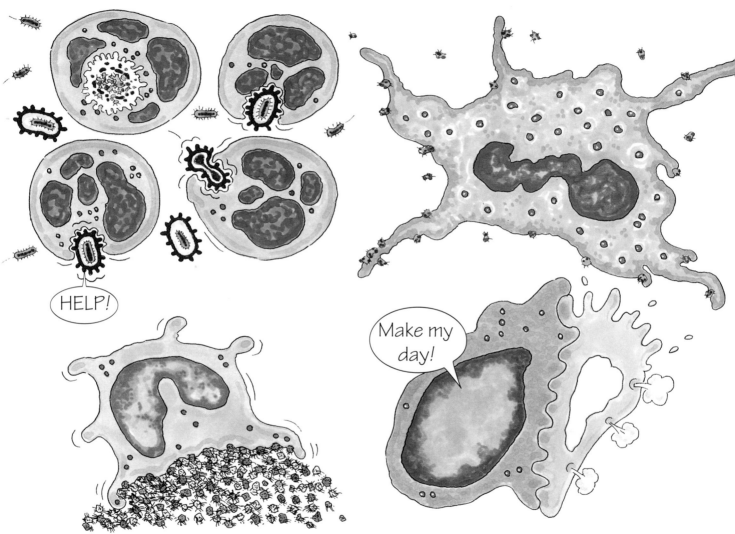

HELP!

Make my day!

Macrophages (mack-row-fages) are garbage disposal experts. They crawl around your body cleaning up the mess when you are injured or ill.

Some **lymphocyte** (lim-foe-site) defender cells make special weapons called antibodies to attack germs. Other lymphocytes are deadly killers that zap sick and dying cells.

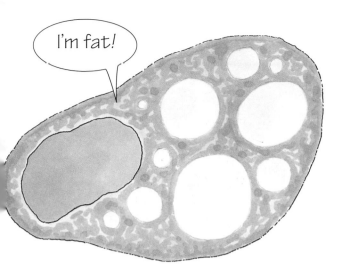

Fat cells are full of... fat! Very important for keeping you warm when it's cold outside. And they make sure you do not "clunk" when you sit down.

Cells called **osteoblasts** (os-tee-oh-blasts) make a protein called collagen. Collagen mixes with minerals to form hard crystals—this is your bone.

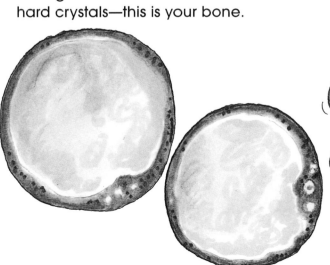

Bone marrow is inside some of your bones. **Stem cells** of your bone marrow divide so quickly that they make 300 million new red blood cells every minute to replace those that die.

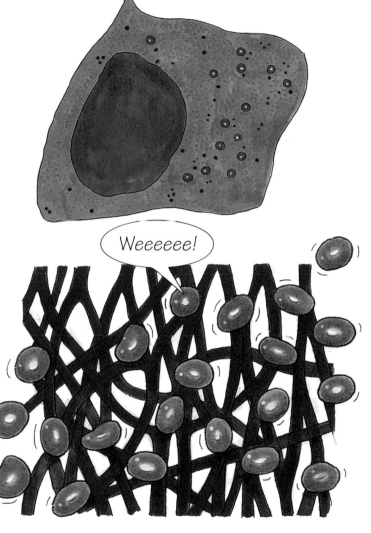

Red blood cells whoosh through your blood tubes. They carry life-giving oxygen to every other cell in your body.

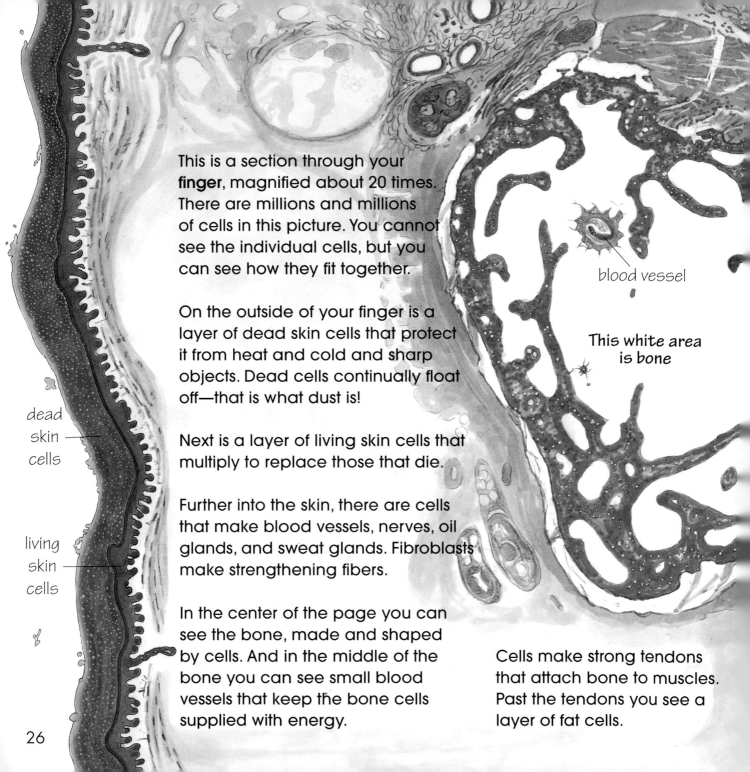

This is a section through your **finger**, magnified about 20 times. There are millions and millions of cells in this picture. You cannot see the individual cells, but you can see how they fit together.

On the outside of your finger is a layer of dead skin cells that protect it from heat and cold and sharp objects. Dead cells continually float off—that is what dust is!

Next is a layer of living skin cells that multiply to replace those that die.

Further into the skin, there are cells that make blood vessels, nerves, oil glands, and sweat glands. Fibroblasts make strengthening fibers.

In the center of the page you can see the bone, made and shaped by cells. And in the middle of the bone you can see small blood vessels that keep the bone cells supplied with energy.

Cells make strong tendons that attach bone to muscles. Past the tendons you see a layer of fat cells.

blood vessel

This white area is bone

dead skin cells

living skin cells

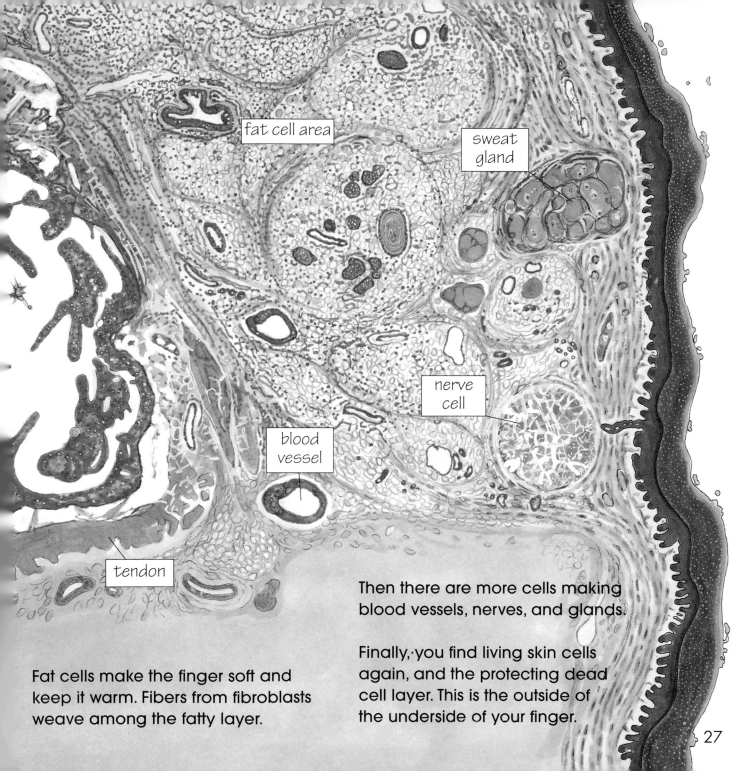

fat cell area

sweat gland

nerve cell

blood vessel

tendon

Then there are more cells making blood vessels, nerves, and glands.

Finally, you find living skin cells again, and the protecting dead cell layer. This is the outside of the underside of your finger.

Fat cells make the finger soft and keep it warm. Fibers from fibroblasts weave among the fatty layer.

27

But what about cells from other plants and animals—are they the same as your cells?

The chemicals that make all cells on planet Earth are very much the same.

All cells have DNA instructions. The DNA code spells different protein messages for each type of animal or plant.

And all cells have a membrane and those complicated bits that we showed you on page 6.

Many of your cells are almost identical to cells that make a dog or cat.

But your cells are rather different from cells that make trees or flowers. Outside the fragile membrane that surrounds a leaf cell, for instance, you will find a rigid wall of cellulose (cell-you-lows) and then a layer of wax. And inside that leaf cell you will see large bubbles of liquid. They help keep the cell, and the whole plant, rigid.

Plant Cell

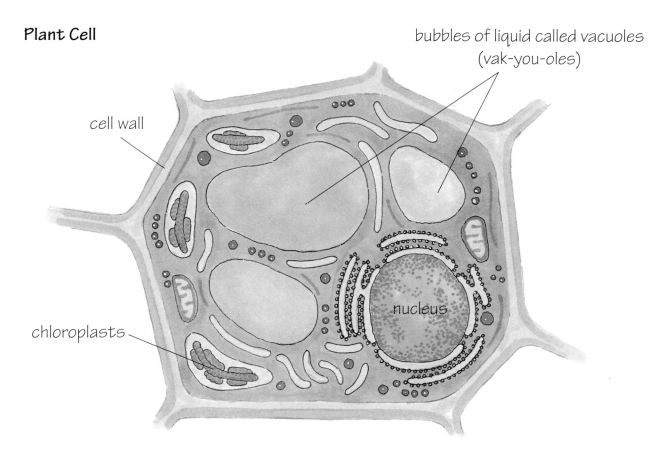

bubbles of liquid called vacuoles (vak-you-oles)

cell wall

chloroplasts

nucleus

There are also objects called chloroplasts (clor-row-plasts) which have a vital job to do. They capture energy from the sun and make it into lovely food for us all to eat.

As we have already told you, your body is made of millions of clever and interesting human cells. But here is a fact that may very well surprise you.

Some parts of your body are covered with cells that are not human at all!

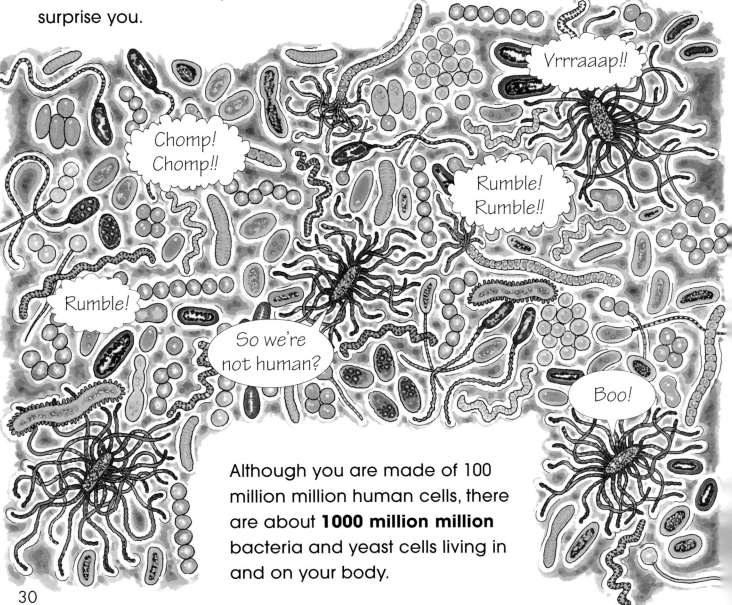

Although you are made of 100 million million human cells, there are about **1000 million million** bacteria and yeast cells living in and on your body.

Do not worry though—they are only there to help us!

Many of your good and helpful bacteria live in your digestive system. They digest the food that you cannot. They also make the useful vitamin K, which helps your blood to clot if you cut yourself.

As the bacteria grow and multiply, they also release gases.